S0-AQJ-811

PERCENTS

Connie Eichhorn
Brian and Janice Wiersema

Power Math Series

CAMBRIDGE Adult Education
A Division of Simon & Schuster
Upper Saddle River, New Jersey

Dr. Connie Eichhorn is the Supervisor of Transitional Programs for the Omaha Public Schools. She is the former president of the American Association of Adult and Continuing Education. Dr. Eichhorn is very active in adult education and has consulted in the development of a variety of adult education materials.

AUTHORS: Brian and Janice Wiersema

EXECUTIVE EDITOR: Mark Moscowitz

EDITOR: Kirsten Richert

PRODUCTION DIRECTOR: Penny Gibson

PRODUCTION EDITOR: Linda Greenberg

PRINT BUYER: Patricia Alvarez

ART DIRECTOR: Marianne Frasco

BOOK DESIGN: Patrice Sheraton

ELECTRONIC PAGE PRODUCTION: Curriculum Concepts

COVER DESIGN: Amy Rosen

COVER PHOTO: © David Bishop/Phototake

Printed in the United States of America

1 2 3 4 5 6 7 8 9 10   99 98 97 96 95

ISBN 0-13-432907-4

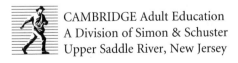
CAMBRIDGE Adult Education
A Division of Simon & Schuster
Upper Saddle River, New Jersey

Contents

To the Learner

The ten books in the Power Math series are designed to help you understand and practice arithmetic skills. Lessons are easy to use and the problems are designed to address every-day adult life.

Lessons have the following features:

- Every lesson begins with a sample problem from real-life experience. You are asked to use your knowledge of math to find a solution.

- The *Think* section takes you through the thought process you might use to organize the information in the problem and choose a problem-solving approach.

- The *Do* section shows you step-by-step how to solve the problem.

- In *Try These*, you will warm up by solving a few problems similar to the opening sample problem. Some steps are worked for you to get you off to a good start.

- The *Practice* section gives you ample opportunities to practice the skill presented in the lesson.

- The *Solving Problems* section applies the math skill in a practical application from your experience as a consumer and worker.

Within each book, review lessons give you opportunities to decide whether you have mastered the skills presented in the book. The *Answer Key* section at the end of the book has answers and worked-out solutions for the problems in the book. Use the answers to check your work. Use the worked-out solutions to make sure your approach to a problem was the correct one.

By working carefully through the exercises in this book, you will find increased confidence in your math skills. Good luck.

PERCENTS

Fractions

A **fraction** is part of a whole or of a group. For example, a dime is one of ten equal parts of a dollar. A penny is one of 100 equal parts of a dollar.

10 dimes = 1 dollar
One dime = $\frac{1}{10}$, "one tenth"

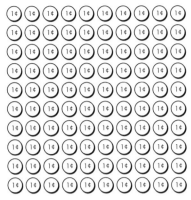

100 pennies = 1 dollar
1 penny = $\frac{1}{100}$, "one hundredth"

When you divide any whole thing into equal parts, the parts can be written as fractions.

1 part of ten parts = $\frac{1}{10}$
"one tenth"

1 part of 100 parts = $\frac{1}{100}$
"one hundredth"

Think

Fractions are used to show equal parts of a whole. A fraction has a numerator and a denominator. The **denominator** is the bottom number and it shows the total number of equal parts in the whole. The top number is the **numerator** and it shows how many equal parts are being shown.

Do

César cut a pizza into eight equal slices. He ate five slices. Write a fraction to show how much he ate.

If you put all the equal parts together, you have the whole thing. All the equal parts of the pizza equal $\frac{8}{8}$.

$$\frac{\text{Numerator}}{\text{Denominator}} = \frac{5}{8} \quad \begin{array}{l} \text{(the number of slices eaten)} \\ \text{(the total number of slices in the pizza)} \end{array}$$

Try These

Write a fraction for the shaded part of each object.

1.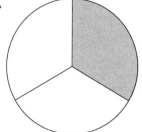

2.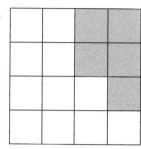

3.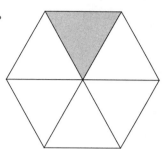

_____   _____   _____

PRACTICE

Write a fraction for each.

4. A circle is cut into five equal parts. Three parts are shaded. Write a fraction for the shaded part. _____

5. A square is cut into four equal parts. One part is shaded. Write a fraction for the shaded part. _____

6. A circle is cut into nine equal parts. Five parts are shaded. Write a fraction for the shaded part. _____

7. Chinua cut the pizza that he ordered for dinner into eight pieces. His wife ate three pieces. Write a fraction for the part of the whole pizza that Chinua ate. _____

8. Joe's company baseball team has 15 players. Two of the players were sick and missed a game. Write a fraction for the part of the team present at the game. _____

9. Sunny and her friends gave out 75 leaflets. If they started with 100 leaflets, write a fraction to show how many they gave out. _____

10. Anna started an exercise program. She plans to work out seven days a week. Write a fraction to show the part of each week she plans to exercise. _____

Solving Problems

Solve.

11. Jim bought five lottery tickets. He won money on two of them. Write a fraction to express the part of his lottery tickets that won money.

    _____

12. Kyong spent $13 on bus tokens out of her weekly budget of $100. Write a fraction to express how much of her weekly budget she spends on bus tokens.

    _____

Check your answers on page 59.

Fractions with the same or equal value are called **equivalent fractions.** Look at these figures.

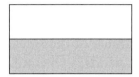

Shaded part = $\frac{1}{2}$ Shaded part = $\frac{2}{4}$ Shaded part = $\frac{4}{8}$

All these fractions have the same value. They are equivalent fractions.

Think

In fractions that have the value of 1, the numerator and the denominator are the same.

Shaded part = $\frac{2}{2}$ Shaded part = $\frac{4}{4}$ Shaded part = $\frac{8}{8}$

$\frac{2}{2}, \frac{4}{4},$ and $\frac{8}{8}$ are equal to 1.

Fractions, like any other number, can be multiplied or divided by 1 without changing their values.

$$5 \times 1 = 5 \qquad \frac{2}{3} \times 1 = \frac{2}{3}$$

$$5 \div 1 = 5 \qquad \frac{2}{3} \div 1 = \frac{2}{3}$$

Multiplying a fraction by a fraction equal to 1 does not change its value.

$$\frac{1}{2} \times \frac{4}{4} = \frac{4}{8}$$

$\frac{1}{2}$ and $\frac{4}{8}$ are equivalent fractions.

Dividing a fraction by a fraction equal to 1 also does not change the value of the first fraction.

$$\frac{4}{8} \div \frac{4}{4} = \frac{1}{2} \qquad \frac{4}{8} = \frac{1}{2}$$

Do

Sometimes you will need to find an equivalent fraction when you are given the numerator or denominator.

$$\frac{1}{2} = \frac{}{6}$$

Step 1. You know the denominator of the equivalent fraction. The first denominator, 2, was multiplied by 3 to get the new denominator, 6.

$$\frac{1}{2 \times 3} = \frac{}{6}$$

Step 2. To find the new numerator multiply the first numerator by 3.

Follow the same steps when dividing.

$$\frac{1 \times 3}{2 \times 3} = \frac{3}{6}$$

$$\frac{8}{12} = \frac{}{3}$$

Try These

Find the missing numbers.

1. $\frac{3}{4} = \frac{6}{}$

2. $\frac{9}{10} = \frac{}{40}$

3. $\frac{16}{20} = \frac{8}{}$

4. $\frac{7}{8} = \frac{}{32}$

5. $\frac{15}{25} = \frac{}{5}$

6. $\frac{9}{12} = \frac{3}{}$

7. $\frac{2}{3} = \frac{8}{}$

8. $\frac{25}{30} = \frac{}{6}$

9. $\frac{4}{7} = \frac{12}{}$

Think

Reducing a Fraction to Lower Terms

Dividing a fraction by 1 written as a fraction is called "reducing a fraction to lower terms" or "simplifying fractions." Dividing $\frac{6}{18}$ by $\frac{6}{6}$ simplifies that fraction to $\frac{1}{3}$, or gives the lower terms, $\frac{1}{3}$.

Do

Reduce $\frac{25}{100}$ to lower terms.

$$\frac{25}{100} \div \frac{25}{25} = \frac{1}{4}$$

$$\frac{25}{100} = \frac{1}{4}$$

Ask yourself: what is the largest number that can divide the numerator and the denominator of $\frac{25}{100}$ evenly.

PRACTICE

Reduce to lower terms.

10. $\frac{50}{100}$ _____

11. $\frac{15}{100}$ _____

12. $\frac{6}{10}$ _____

13. $\frac{12}{50}$ _____

14. $\frac{10}{100}$ _____

15. $\frac{75}{100}$ _____

Think

Raising a Fraction to Higher Terms

Multiplying a fraction by 1 written as a fraction is called "raising to higher terms."

Do

Raise $\frac{4}{5}$ to hundredths.

$$\frac{4}{5} \times \frac{20}{20} = \frac{80}{100}$$

$$\frac{4}{5} = \frac{80}{100}$$

What form of 1 written as a fraction can you use to find this out? Ask yourself: What fraction can you multiply by the numerator and denominator to raise that fraction to hundredths?

Raise to higher terms.

16. $\frac{1}{4}$ to hundredths _____

17. $\frac{3}{5}$ to tenths _____

18. $\frac{7}{20}$ to hundredths _____

19. $\frac{1}{10}$ to hundredths _____

20. $\frac{2}{12}$ to tenths _____

21. $\frac{9}{50}$ to hundredths _____

Solving Problems

Solve.

22. Jason, a newspaper reporter, is writing a story about highway safety. He finds that during the month of July a total of 75 traffic tickets were issued for speeding. He also discovers that 50 of the tickets were given to out-of-state cars. The fraction $\frac{50}{75}$ represents the number of out-of-state cars receiving tickets. Reduce $\frac{50}{75}$ to lower terms.

_____

23. Nikki read a magazine article that said the average person spends about $\frac{1}{3}$ of every 24 hours sleeping. Nikki always tries to get 8 hours of sleep each night. She wonders if she was above or below the average.

The part of each day (24 hours) she spends asleep can be expressed by the fraction $\frac{8}{24}$. Is $\frac{8}{24}$ greater than, less than, or equal to $\frac{1}{3}$?

_____

Check your answers on page 59.

Improper Fractions and Mixed Numbers

Fractions that are greater than or equal to 1 are called improper fractions. In an improper fraction the numerator is greater than or equal to the denominator.

$\frac{7}{3}$ and $\frac{20}{5}$ are improper fractions.

A whole and a fraction together, such as $3\frac{1}{4}$ or $1\frac{1}{2}$, is called a mixed number.

Think

Mixed numbers can be expressed as improper fractions. Improper fractions can be expressed as mixed numbers

Mixed Number

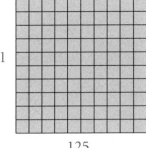

1

$\frac{125}{100}$

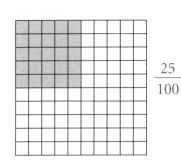

$\frac{25}{100}$

Improper fraction

Do

Change $2\frac{3}{4}$ to an improper fraction.

To change a mixed number to an improper fraction:

Step 1. Multiply the whole number by the denominator. $2 \times 4 = 8$

Step 2. Add the product and the numerator. $8 + 3 = 11$

Step 3. Write the sum over the denominator. $\frac{11}{4}$ $2\frac{3}{4} = \frac{11}{4}$

Try These

Write each mixed number as an improper fraction.

1. $5\dfrac{3}{5} =$ _____

2. $2\dfrac{1}{2} =$ _____

3. $3\dfrac{2}{3} =$ _____

4. $4\dfrac{7}{8} =$ _____

5. $7\dfrac{1}{4} =$ _____

6. $8\dfrac{1}{3} =$ _____

7. $5\dfrac{3}{8} =$ _____

8. $1\dfrac{3}{7} =$ _____

Think

Improper fractions can be expressed as mixed numbers.

Do

Change $\frac{125}{100}$ to a mixed number.

To change an improper fraction to mixed number:

Step 1. Divide the numerator by the denominator.

$$100\overline{)125}^{\,1\frac{25}{100}}$$
$$\underline{100}$$
$$25$$

Step 2. Write the remainder as a fraction. Reduce the fraction to lower terms.

$$1\dfrac{25}{100} = 1\dfrac{25}{100} \div \dfrac{25}{25} = 1\dfrac{1}{4}$$

Try These

Write each improper fraction as a mixed number.

9. $\dfrac{7}{3} =$ _____

10. $\dfrac{9}{2} =$ _____

11. $\dfrac{20}{6} =$ _____

12. $\dfrac{5}{3} =$ _____

13. $\dfrac{150}{100} =$ _____

14. $\dfrac{12}{10} =$ _____

15. $\dfrac{75}{50} =$ _____

16. $\dfrac{15}{8} =$ _____

Check your answers on page 59.

LESSON

4

Decimals

Decimals, like fractions, show how a whole or a group is divided into equal parts. The figures below are divided into 10 equal parts and each part equals $\frac{1}{10}$ or one tenth. One tenth as a decimal is written as .1, three tenths is written as .3, and .6 is the decimal for six tenths.

Think

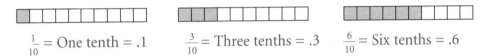

$\frac{1}{10}$ = One tenth = .1 $\frac{3}{10}$ = Three tenths = .3 $\frac{6}{10}$ = Six tenths = .6

Decimals, like fractions, can be shown in hundredths.

Do

The squares are divided into 100 equal parts, or hundredths. Hundredths are written as two-place decimals.

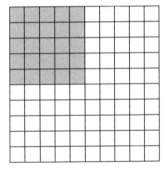
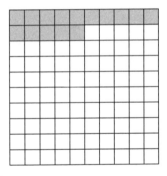

$\frac{25}{100}$ = Twenty-five hundredths = .25 $\frac{15}{100}$ = Fifteen hundredths = .15

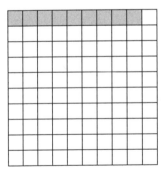

$\frac{9}{100}$ = Nine hundredths = .09

With decimals, as with fractions, all the equal parts total the whole amount or 1.

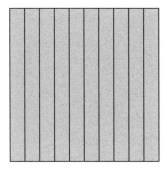

$\frac{10}{10}$ = Ten tenths = 1.0

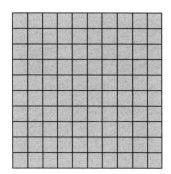

$\frac{100}{100}$ = One hundred hundredths = 1.00

Try These

Write the fractions as decimals.

1. $\dfrac{4}{10}$ = _____

2. $\dfrac{5}{100}$ = _____

3. $\dfrac{35}{100}$ = _____

4. $\dfrac{8}{10}$ = _____

5. $\dfrac{7}{10}$ = _____

6. $\dfrac{45}{100}$ = _____

7. $\dfrac{18}{100}$ = _____

Think

When reading and using whole numbers, every number has a place value.

You read decimals the same way; you must know the place value of each digit.

Read a decimal like a whole number. When you get to the decimal point say "and," and then read the rest of the number. Finish by saying the place value of the last digit.

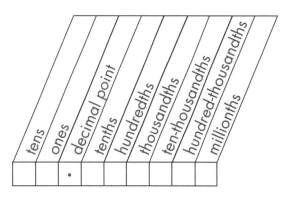

Do

Read each decimal.

34.6	thirty-four **and** six **tenths**
2.18	two **and** eighteen **hundredths**
.807	eight hundred seven **thousandths**
.09	nine **hundredths**
11.25	eleven **and** twenty-five **hundredths**
.7462	seven thousand, four hundred sixty-two ten **thousandths**

Try These

Write each decimal number in words.

8. 18.08   eighteen and eight hundredths

9. 2.2   _____

10. .1856   _____

11. .35   _____

12. 4.91   _____

PRACTICE

Write each as decimals.

13. six and nine hundredths _____

14. five hundred one and four tenths _____

15. twenty-five and four hundred two thousandths _____

16. one thousand fifty-six and twenty-two hundredths _____

17. nine hundred sixty-three and eight tenths _____

Check your answers on page 60.

5

Changing Fractions to Decimals and Decimals to Fractions

You use fractions and decimals to solve percent problems. Changing fractions into decimals or decimals into fractions will help you solve these problems.

Think

Changing Fractions to Decimals

You already know that both fractions and decimals are used to show parts of a whole or parts of a group. Many decimals have a value of tenths and hundredths. When a fraction has a denominator of 10, 100, 1000 and so on, it can be changed easily into a decimal.

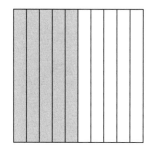

$\frac{5}{10}$ = Five tenths = .5

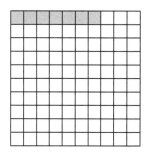

$\frac{7}{100}$ = Seven hundredths = .07

The zero is a place holder.

You can raise fractions to higher terms with denominators of 10, 100, 1000 and so on.

$$\frac{1}{2} \times \frac{5}{5} = \frac{5}{10} = .5$$

$$\frac{1}{4} \times \frac{25}{25} = \frac{25}{100} = .25$$

But what about a fraction like $\frac{5}{6}$? To change this fraction to a decimal, divide the numerator by the denominator.

```
      .833 Rounded off = .83
  6)5.000
    4 8
     20
     18
     20
     18
```

Do

Change $\frac{3}{5}$ to a decimal. Divide the numerator by the denominator.

$$\begin{array}{r} .6 \\ 5\overline{)3.0} \\ \underline{3\ 0} \end{array}$$

Try These

Change the fractions to decimals.

1. $\frac{12}{100}$ _____

2. $\frac{25}{100}$ _____

3. $\frac{7}{10}$ _____

4. $\frac{9}{1000}$ _____

5. $\frac{3}{5}$ _____

6. $\frac{13}{20}$ _____

PRACTICE

Change the fractions to decimals. Round the answer to the nearest hundredth.

7. $\frac{4}{7}$ _____

8. $\frac{5}{8}$ _____

9. $\frac{4}{5}$ _____

10. $\frac{5}{6}$ _____

Think

Changing Decimals to Fractions

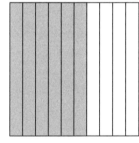

.6 = Six tenths = $\frac{6}{10}$

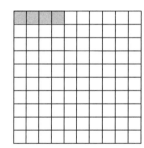

.04 = Four hundredths = $\frac{4}{100}$

> **REMEMBER**
>
> Most decimals you use are tenths and hundredths and are equal to fractions with tenths and hundredths as the denominator.

To change a decimal to a fraction, remember these "Three Rs."

Step 1. Read it!

Step 2. (w)Rite it!

Step 3. Reduce it!

Do

Change .15 to a fraction.

Step 1. Read it.   fifteen hundredths

Step 2. (w)Rite it. $\dfrac{15}{100}$

Step 3. Reduce it. $\dfrac{15}{100} = \dfrac{3}{20}$

Try These

Change each decimal to a fraction.

11. .6 = _____

12. .45 = _____

13. .25 = _____

14. .08 = _____

15. .12 = _____

16. .80 = _____

Solving Problems

Solve.

17. Pat wants to make a meatloaf. The recipe calls for $\frac{3}{4}$ of a pound of hamburger and $\frac{2}{5}$ of a pound of ground pork. She needs to change $\frac{3}{4}$ and $\frac{2}{5}$ to decimals so she can buy the right amounts of meat to make the meatloaf.

_____

18. Jomo is buying food for a picnic. His girlfriend said to buy about $\frac{3}{4}$ pound of hamburger and $\frac{1}{2}$ pound of potato salad. Change $\frac{3}{4}$ and to $\frac{1}{2}$ decimals.

_____

Check your answers on page 60.

Applying Your Knowledge: Fractions and Decimals

Use what you know about fractions, equivalent fractions, and decimals to complete these activities.

REMEMBER

Multiplying or dividing a fraction by 1 or a form of 1 ($\frac{2}{2}$, $\frac{5}{5}$) does not change its value.

Match each decimal in the first column with the correct fraction and with the correct picture.

Decimal	Fraction	Picture
1. .75	$\frac{1}{5}$	
2. .26	$\frac{1}{4}$	
3. .2	$\frac{3}{10}$	
4. .25	$\frac{13}{50}$	
5. .3	$\frac{3}{4}$	

Match each fraction with the decimal that has the same value.

Fraction	Decimal
6. $\dfrac{1}{5}$	.25
7. $\dfrac{2}{100}$	.2
8. $\dfrac{4}{100}$	.04
9. $\dfrac{5}{10}$	.02
10. $\dfrac{1}{4}$	.5

Solving Problems

Solve.

11. Maggie baked a cherry pie for supper. As she was getting ready for bed, she opened the refrigerator to get a drink of water. She saw the pie pan sitting on the top shelf. It looked like this.

    Write a fraction to represent how much of the pie was left. Also, express it as a decimal. Can you think of another fraction that would have the same value?

_____

Check your answers on page 60.

7

What Is Percent?

A percent, like a fraction and a decimal, is also a way to describe the relationship of a part to a whole. A percent always describes the relationship to a whole divided into 100 parts. The word *percent* means "for each 100" or "out of 100." The sign for percent is %. The whole is 100%.

Think

The square on the right is divided into 100 equal parts. 50 of the parts are shaded. The shaded portion is 50 out of 100 or 50% of the whole.

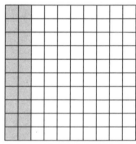

Do

Look at this square. Out of its 100 equal parts, 20 are shaded and 80 are unshaded. What percent of the square is shaded? _20_% What percent of the square is unshaded? _80_%

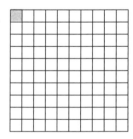

Try These

For problems 1 to 9, tell what percent of each grid is shaded.

1.

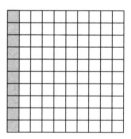

_____ % shaded

2.

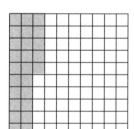

_____ % shaded

3.

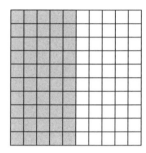

_____ % shaded

4.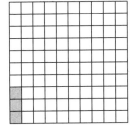

_____ % shaded

5.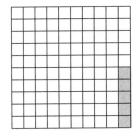

_____ % shaded

6.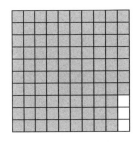

_____ % shaded

7.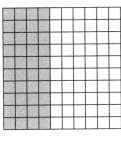

_____ % shaded

8.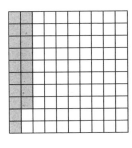

_____ % shaded

9.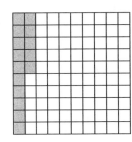

_____ % shaded

PRACTICE

For problems 10 to 15, shade the correct percent of each grid.

10.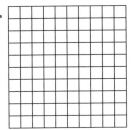

shade 20%

11.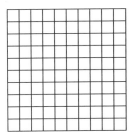

shade 60%

12.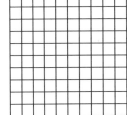

shade 35%

13.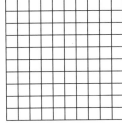

shade 2%

14.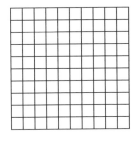

shade 51%

15.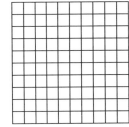

shade 99%

Check your answers on page 61.

Sales at Mondo Pizza, Inc. increased 150% when the company started delivery service. If 100% means the whole thing, what does 150% mean?

Think

Some percents show more than a whole amount. Look at the squares below. The first square shows that 100% of the square is shaded. The second square shows 50% more is shaded.

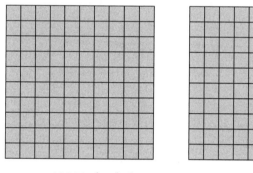

100% shaded   +   50% shaded

Add the percents. 150% of the whole amount is shaded.

Do

In the problem above, Mondo Pizza's sales this year are greater than the sales last year. So they sold 100% of last year's sales and 50% *more*, or 150%.

$$100\% + 50\% = 150\%$$

Try These

What percent of the squares are shaded?

1.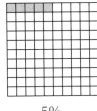

100%   5%

shaded portion _____

2.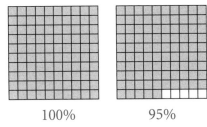

100%   95%

shaded portion _____

3.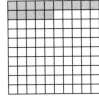

shaded portion _____

4.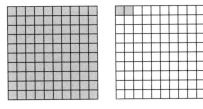

shaded portion _____

5.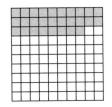

shaded portion _____

6.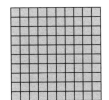

shaded portion _____

7.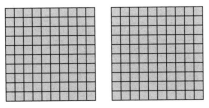

shaded portion _____

8.

shaded portion _____

9.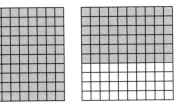

shaded portion _____

10. If 100% represents one square, draw a picture that would show 175%.

11. If 100% represents one apple, draw a picture that would show 200%.

12. If 4 = 100%, then 8 = _____ %

13. If 6 = 100%, then 9 = _____ %

Solving Problems

Solve.

14. Over the past two years, the number of people in the exercise program at the community center has increased 300%. Draw squares to show this percent.

15. Because of inflation, the price of beef went up 150%. What does this mean?

_____

_____

Check your answers on page 61.

Applying Your Knowledge: Meaning of Percent

This lesson will help you practice what you have learned about percents.

1. Percent means _____.

2. A percent can be expressed as a _____ or as a

   _____.

For each square, write the percent that is shaded and the percent that is not shaded.

3.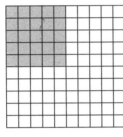

_____ shaded

_____ not shaded

4.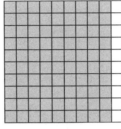

_____ shaded

_____ not shaded

5.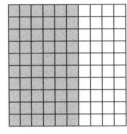

_____ shaded

_____ not shaded

6.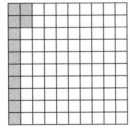

_____ shaded

_____ not shaded

7.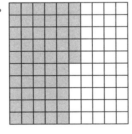

_____ shaded

_____ not shaded

8.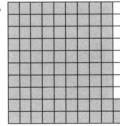

_____ shaded

_____ not shaded

For the following figures, give the percent of shaded portions in the two squares together.

9.

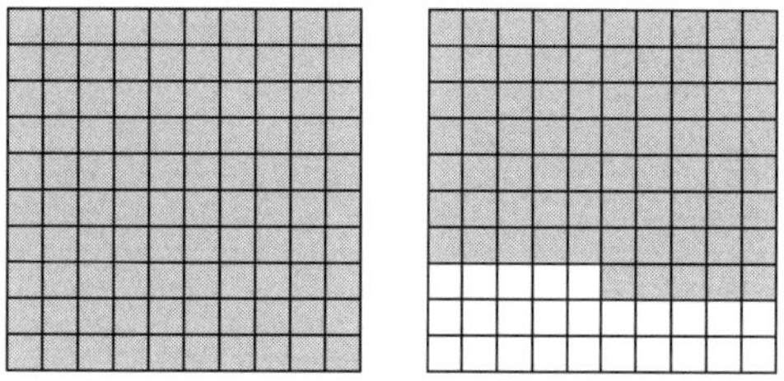

shaded portion _____

10.

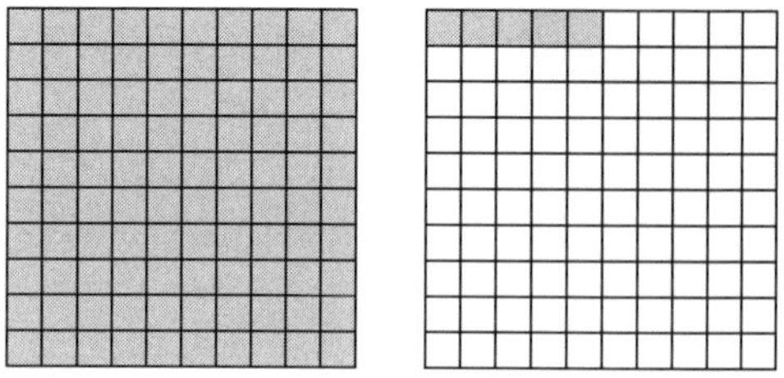

shaded portion _____

11.

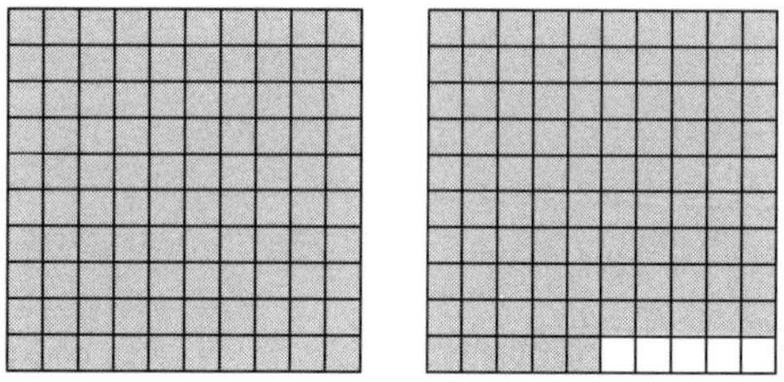

shaded portion _____

12.

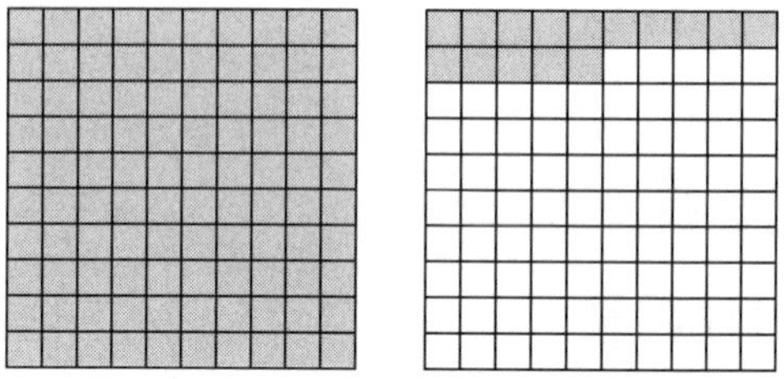

shaded portion _____

13.

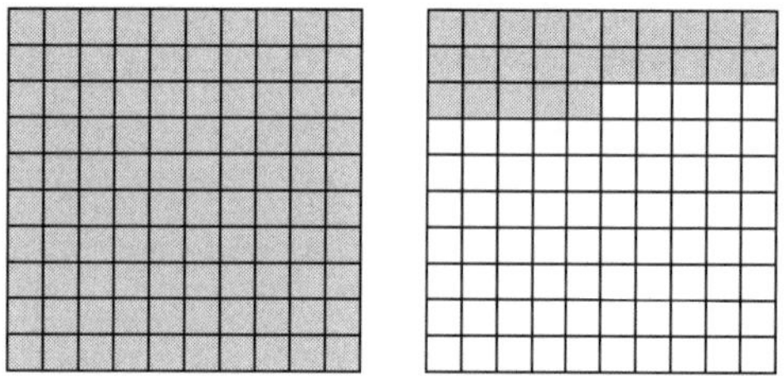

shaded portion _____

14.

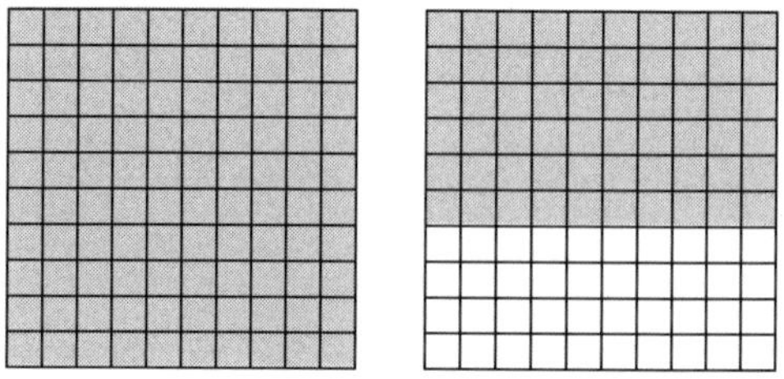

shaded portion _____

Solving Problems

Solve.

15. Victor spent 2% of his salary on lottery tickets this month. How many dollars out of every 100 dollars he earns did he spend on the lottery?

_____

16. The rent on Ono's apartment went up 125% over the past five years. Draw a picture to represent this percent.

_____

17. Ann and Carlos, two part-time employees of the Ajax Copying Company, asked for a 5% raise. Write this percent as a fraction. Reduce the fraction.

_____

18. George sells used cars. He expected to receive six used cars today. Only 50% of the people with used cars brought them in, however. Draw a picture to represent the percent of used cars George received.

_____

19. If 72% of the employees at the discount clothing outlet in the Mahasett mall have health coverage, what percent do not have health coverage?

_____

20. In Chicago, 5,000 taxpayers are going to get back money from the state government because they paid too much tax in April. As of early September, 0% of them have received their tax refund. How many of the 5,000 have received their tax refund?

_____

Check your answers on page 61.

LESSON

10

Changing Percents to Fractions

Tadashi is in charge of buying food for a block association street fair. He found that 75% of the people want chicken and 25% want beef. Change 75% and 25% to fractions.

All percents can be written as fractions with 100 as the denominator. Drop the percent sign. The number becomes the numerator of the fraction. The denominator is 100. Reduce the fraction if needed.

Do

Change 25% to a fraction.

Step 1. Drop the percent sign. Write the number over 100. $\dfrac{25}{100}$

Step 2. Reduce.

$$\frac{25}{100} \div \frac{25}{25} = \frac{1}{4} \qquad\qquad 25\% = \frac{1}{4}$$

Change 75% to a fraction.

Step 1. $75\% = \dfrac{75}{100}$

Step 2. $\dfrac{75}{100} \div \dfrac{25}{25} = \dfrac{3}{4}$ $\qquad\qquad 75\% = \dfrac{3}{4}$

Tadashi bought chicken for $\frac{3}{4}$ of the people and beef for $\frac{1}{4}$ of the people.

Try These

Write each percent as a fraction in simplest form.

1. $17\% = \dfrac{}{100}$

2. $80\% = \dfrac{}{100}$

3. $20\% = $ _____

4. $32\% = $ _____

Write each percent as a fraction in its simplest form.

5. 50% = _____   **6.** 28% = _____   **7.** 35% = _____   **8.** 68% = _____

9. 15% = _____   **10.** 56% = _____   **11.** 78% = _____   **12.** 62% = _____

13. 90% = _____   **14.** 5% = _____   **15.** 48% = _____   **16.** 84% = _____

Solving Problems

Change the percent to a fraction and reduce.

17. At a factory, 80% of the union workers voted to accept the new contract offered by their employer. Write a fraction for the number of workers voting for the contract.

18. At the pet store where Karl works, a customer ordered a special cat food mixture. The mixture contains 65% organic grains. Write a fraction to show how much of each bag of the mixture contains organic grains.

_____   _____

Check your answers on page 61-62.

Changing Fractions to Percent

José answered $\frac{3}{4}$ of the questions on his driver's test correctly. José was told that he must achieve a score above 70% in order to receive his license. Did José score high enough on his test to receive his license?

Think

To change a fraction to a percent, first change the fraction to a decimal. Divide the numerator by the denominator. Then move the decimal point two places to the right and add a percent sign.

Do

Change $\frac{3}{4}$ to a percent.

Step 1. Divide 3 by 4.

$$\begin{array}{r} .75 \\ 4\overline{)3.00} \\ \underline{2\ 8} \\ 20 \\ \underline{20} \end{array}$$

Step 2. Move the decimal point two places to the right.

.75

Step 3. Add a percent sign.

75%

José got 75% of the questions correct, which is more than the 70% needed. He will receive his license!

Try These

Change each fraction to a percent.

1. $\frac{1}{2} = 2\overline{)1.0} = $ _____ %   2. $\frac{7}{10} = $ _____ %   3. $\frac{9}{20} = $ _____ %

4. $\frac{1}{4} = $ _____ %   5. $\frac{7}{20} = $ _____ %   6. $\frac{3}{5} = $ _____ %

Do

Some fractions will have a remainder after dividing. Divide to three decimal places (thousandths), then round the answer to hundredths. Move the decimal point two places to the right and add the percent sign.

$$\frac{1}{8} = \begin{array}{r} .125 \\ 8\overline{)1.000} \\ \underline{8} \\ 20 \\ \underline{16} \\ 40 \\ \underline{40} \end{array}$$

.125 rounds to .13

.13 = 13%

PRACTICE

7. $\dfrac{1}{8} =$ _____

8. $\dfrac{5}{6} =$ _____

9. $\dfrac{11}{20} =$ _____

10. $\dfrac{2}{3} =$ _____

11. $\dfrac{17}{100} =$ _____

12. $\dfrac{7}{9} =$ _____

13. $\dfrac{8}{10} =$ _____

14. $\dfrac{9}{25} =$ _____

Solving Problems

Solve.

15. Roberto and Julia went to watch their son Julio play football. Roberto kept track and figured that his son completed 13 out of 20 passes. What percent of his passes did he complete?

_____

16. Amal wants to buy a used car. He paid $\frac{3}{20}$ of the total price as a down payment. Write this as a percent.

_____

Check your answers on page 62.

12

Applying Your Knowledge: Percents and Fractions

Match the fraction in the first column with the correct percent and with the correct picture.

Fraction	Percent	Picture
1. $\frac{7}{20}$	50%	
2. $\frac{1}{5}$	60%	
3. $\frac{3}{10}$	15%	
4. $\frac{1}{2}$	75%	
5. $\frac{3}{20}$	30%	
6. $\frac{3}{4}$	20%	
7. $\frac{3}{5}$	35%	

Complete the following.

Fraction	Percent	Picture

8. _____ 90%

9. $\dfrac{1}{4}$ _____

10. _____ 65%

11. $\dfrac{2}{5}$ _____

12. _____ 80%

13. $\dfrac{}{10}$ _____

14. _____ 45%

Write the fraction and the percent that each picture represents.

	Fraction	Percent

15.

_____   _____

16.

_____   _____

Check your answers on page 62-63.

50% and 25%

Jorge's uncle told him that most cars that are a year old are worth only 50% of their original price. What is another way to think about 50%?

Think

The figure below shows that 50 squares out of 100 are shaded. So, 50% are shaded.

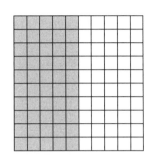

50% shaded
$\frac{1}{2}$ is shaded

You can also say that $\frac{50}{100}$ are shaded.

$$\frac{50}{100} = \frac{5}{10} = \frac{1}{2}$$

So, Jorge's uncle could have said, "Most cars a year old are only worth $\frac{1}{2}$ of their original price."

If you think of common percents as fractions it will help you solve percent problems.

Do

Let's think about 25%. 25% means 25 out of 100 parts. In the grid below, 25 parts out of 100 parts are shaded.

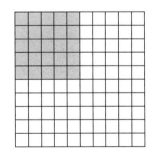

25% is shaded
$\frac{1}{4}$ is shaded

Write 25% as a fraction, $\frac{25}{100}$.

$$\frac{25}{100} \div \frac{25}{25} = \frac{1}{4}$$

When you see 25%, think of $\frac{1}{4}$.

Try These

Shade 50% or $\frac{1}{2}$ of the figures in each problem.

1.  **2.**  **3.**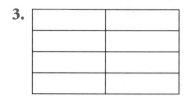

Shade 25% or $\frac{1}{4}$ of the figure in each problem.

4. 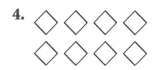 **5.** 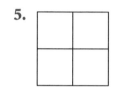 **6.**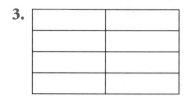

Check your answers on page 63.

Common Percents and Fraction Equivalents

The table below shows common percent and fraction equivalents. Study the table to think about fractions as percents and percents as fractions.

COMMON EQUIVALENTS

	Fraction	Percent
Twentieths	$\frac{1}{20}$	5%
Tenths	$\frac{1}{10}$	10%
	$\frac{3}{10}$	30%
	$\frac{7}{10}$	70%
	$\frac{9}{10}$	90%
Eighths	$\frac{1}{8}$	$12\frac{1}{2}$%
	$\frac{3}{8}$	$37\frac{1}{2}$%
	$\frac{5}{8}$	$62\frac{1}{2}$%
	$\frac{7}{8}$	$87\frac{1}{2}$%
Sixths	$\frac{1}{6}$	$16\frac{2}{3}$%
	$\frac{5}{6}$	$83\frac{1}{3}$%
Fifths	$\frac{1}{5}$	20%
	$\frac{2}{5}$	40%
	$\frac{3}{5}$	60%
	$\frac{4}{5}$	80%
Quarters	$\frac{1}{4}$	25%
	$\frac{3}{4}$	75%
Thirds	$\frac{1}{3}$	$33\frac{1}{3}$%
	$\frac{2}{3}$	$66\frac{2}{3}$%
Half	$\frac{1}{2}$	50%

Use the table to complete each problem.

1. $50\% =$ _____

2. $\dfrac{1}{5} =$ _____

3. $12\dfrac{1}{2}\% =$ _____

4. $\dfrac{1}{3} =$ _____

5. $10\% =$ _____

6. $\dfrac{3}{4} =$ _____

7. $5\% =$ _____

8. $\dfrac{1}{4} =$ _____

9. $33\dfrac{1}{3}\% =$ _____

10. $\dfrac{1}{2} =$ _____

11. $25\% =$ _____

12. $\dfrac{1}{10} =$ _____

13. $66\dfrac{2}{3}\% =$ _____

14. $\dfrac{1}{8} =$ _____

15. $75\% =$ _____

16. $\dfrac{1}{20} =$ _____

17. $20\% =$ _____

18. $\dfrac{2}{3} =$ _____

Solving Problems

Solve.

19. All summer merchandise at Jean's is on sale for 50% off the regular price starting at 8:00 tomorrow morning. Express the percent as a fraction.

_____

20. The strike started at midnight and $\frac{3}{5}$ of the workers walked out. What percent of the workers walked out?

_____

21. Joshua is studying for the test to get his motorcycle license. He needs at least an 80% on the test. What fraction of the questions must he get right in order to get his license?

_____

22. Mr. Rivera agreed to help his son Ernesto buy a bike. If Ernesto saves enough for two fifths of the total cost, his dad will pay the rest. What percent of the bike must Ernesto be able to pay for? What percent is Mr. Rivera willing to pay?

_____

23. After paying the bills last month, Brenda figured that $\frac{3}{10}$ of the family budget had been spent on groceries. What percent of the budget went to pay for groceries?

24. According to early reports on the television, Nehru's favorite candidate was leading after 5% of the votes were in. Write the percent as a fraction.

_____   _____

Check your answers on page 63.

15

Changing Percents to Decimals

Ramona wanted 6% of her weekly income put in the company savings plan. To find out the amount, she first changed 6% to a decimal.

Think ....

To solve percent problems you often multiply or divide. You often need to change the percent to a decimal.

REMEMBER

Percent means hundredths.

Do

Step 1. To change a 6% to a decimal move the decimal point two places to the left. If the percent is less than 10%, you still move the decimal point two places to the left. You add a zero or zeroes as placeholders.

.06.%

Step 2. Drop the percent sign.

.06

6% = .06

Try These

Write the following percents as decimals.

1. 34% = _____

2. 72% = _____

3. 8% = _____

4. 3% = _____

5. 30% = _____

6. 80% = _____

─────────────── **PRACTICE** ───────────────

Change each percent to a decimal.

7. 7.6% = _____

8. 45.7% = _____

9. 85.5% = _____

10. 3% = _____

11. 10.5% = _____

12. 4.4% = _____

13. 5% = _____

14. 7.25% = _____

Solving Problems

Solve.

15. Randall has to pay 7% interest on a loan. Write the interest rate as a decimal.

16. Jamal pays 8.5% of his income for health insurance. Write the percent as a decimal.

_____

_____

Check your answers on page 64.

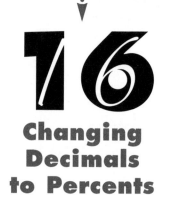

LESSON

16

Changing Decimals to Percents

Jason is a quarterback on a semipro football team. He completed .54 of his passes. What percent of his passes did he complete?

Change .54 to a percent.

Think

Percent means hundredths, so you move the decimal point to change decimals to percents.

Do

Step 1. To change .54 to a percent, move the decimal point two places to the right.

54.

Step 2. Add the percent sign. It is not necessary to write the decimal point. It is understood to be to the right of the number.

54%

If there is only one decimal place, add a zero as a place holder.

.3 = 30. = 30%

If there are more than two decimal places, you must keep the decimal point.

.326 = 32.6% = 32.6%

Try These

Change the decimals to percents.

1. .63 = _____   2. .51 = _____   3. .2 = _____

4. .465 = _____   5. .18 = _____   6. .8 = _____

Change the decimals to percents.

7. .49 = _____   **8.** .05 = _____   **9.** .005 = _____   **10.** .6 = _____

11. .281 = _____   **12.** .9 = _____   **13.** .33 = _____   **14.** .876 = _____

15. .2 = _____   **16.** .444 = _____

Solving Problems

Solve.

17. Mikita knows that batting averages are really decimals carried out to the thousandths place. According to the sports page, her favorite player has a batting average of .308. Write the batting average as a percent.

18. Joe wants to increase sales by .15. Write the decimal as a percent.

_____   _____

Check your answers on page 64.

17

Percents Less Than 1%

Mark works in a machine factory. There is a new program called quality control. Management wants to reduce mistakes by .5%.

Change .5% to a fraction and a decimal.

Think

1 percent is the same as one hundredth. If a percent is less than 1 percent it is less than a hundredth. Look at the figures below.

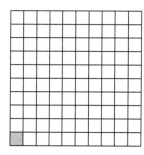

shaded portion 1%

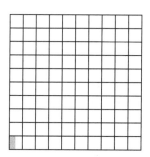

shaded portion .5%

Do

To change .5% into a decimal, move the decimal point two places to the left. Drop the percent sign.

$$.5\% = .005$$

Read the decimal .005 as "five thousandths."

Change it to a fraction.

$$.5\% = .005 = \frac{5}{1000} \div \frac{5}{5} = \frac{1}{200}$$

$$.5\% = .005 = \frac{1}{200}$$

Try These

Change the following percents to decimals and fractions.

1. .3% = .003 = _____

2. .4% _____

3. .25% _____

4. .2% _____

Change the following percents to decimals.

5. .8% = _____

6. 7.5% = _____

7. .15% = _____

8. .098% = _____

9. 3.15% = _____

10. .54% = _____

Change the following percents to fractions.

11. 32% = _____

12. .64% = _____

13. 2.5% = _____

14. .6% = _____

15. .48% = _____

Change the following to percents.

16. .33 = _____

17. .009 = _____

18. $\dfrac{3}{400}$ = _____

19. $\dfrac{3}{250}$ = _____

20. .0039 = _____

21. .0705 = _____

Solving Problems

Solve.

22. The percent of error in the software accounting program is .04%. Write this as a decimal and a fraction.

_____

Check your answers on page 64.

LESSON

18

Decimals and Percents Greater Than 100%

Abbas and his wife are about to have their second child. They need to find a three-bedroom apartment. The rent will be 175% of what they are paying now. Change 175% into a decimal.

Think

Since 175% is greater than 100%, the decimal will be greater than 1.

Do

Change a percent to a decimal.

Step 1. Move the decimal place two places to the left.   1.75%

Step 2. Drop the percent sign.   1.75

175% = 1.75

Try These

Change the percents to decimals.

1. 225% = _____   **2.** 133% = _____   **3.** 315% = _____

4. 185% = _____   **5.** 378% = _____   **6.** 205% = _____

7. 355% = _____   **8.** 450% = _____   **9.** 125% = _____

Do

Change 1.55 to a percent.

To change a decimal to a percent:

Step 1. Move the decimal place two places to the right.   155.

Step 2. Add a percent sign.   155%

1.55 = 155%

Try These

Change the decimals to percents.

10. 1.55 = _____

11. 2.85 = _____

12. 2.35 = _____

13. 3.2 = _____

14. 8.1 = _____

15. 3.05 = _____

16. 4.05 = _____

17. 2.15 = _____

18. 6.01 = _____

PRACTICE

Change to percents or decimals.

19. 256% = _____

20. 1.05 = _____

21. 196% = _____

22. 2.9 = _____

23. 651% = _____

24. 3.75 = _____

25. 758% = _____

26. 4.62 = _____

27. 105% = _____

28. 5.33 = _____

29. 247% = _____

30. 1.08 = _____

Solving Problems

Solve.

31. The profits for Chin's car repair shop have increased 225% since his nephew started working full time. Write this percent as a decimal.

32. The local pharmacy uses a 125% markup to price their prescription drugs. Write the decimal for the markup.

_____

_____

33. The Gators basketball team went from last place to first place. Attendance at home games has improved 300%. Write this percent as a decimal.

_____

34. Property taxes for Raoul Santore's repair shop increased over 220% in the past five years. Write the percent as a decimal.

_____

35. Health insurance costs for the company where Margarita works have increased 115% during this past year. Write the increase as a decimal.

_____

Check your answers on page 65.

19

Changing Percents Greater Than 100% to Mixed Numbers

During a cold winter the amount of heating fuel used increased by 275%. Before Ali figures out the cost of fuel, he changes the percent to a fraction.

Change 275% to a mixed number.

Think

REMEMBER

Percent means hundredths. Since the percent is more than 100%, the fraction will be a mixed number.

Do

Step 1. To change 275% to a mixed number, write the number over 100.

$$\frac{275}{100}$$

Step 2. Reduce.

$$\frac{275}{100} \div \frac{25}{25} = \frac{11}{4}$$

Step 3. Change the improper fraction to a mixed number.

$$275\% = 2\frac{3}{4}$$

Try These

Change the following percents to mixed numbers. Reduce fractions.

1. 250% = _____

2. 125% = _____

3. 375% = _____

4. 220% = _____

5. 180% = _____

6. 425% = _____

7. 330% = _____

8. 115% = _____

Change the following percents to fractions. Change all improper fractions to mixed numbers.

9. 385% = _____   **10.** 215% = _____   **11.** 145% = _____

12. 60% = _____   **13.** 480% = _____   **14.** 105% = _____

15. 5% = _____   **16.** 280% = _____   **17.** 30% = _____

18. 360% = _____   **19.** 390% = _____   **20.** 155% = _____

Solving Problems

Solve.

21. The police have reported an increase of 130% in the number of stolen cars this year. Write 130% as a mixed number.

22. The doctor's office has been crowded recently due to a 225% increase in the number of patients since the flu outbreak. Write 225% as a mixed number.

_____   _____

Check your answers on page 65.

Changing Mixed Numbers to Percents

Will gave away calculators to get more people into his store. He found that $2\frac{1}{2}$ times as many people came today as yesterday.

Change $2\frac{1}{2}$ to a percent.

Think

First, change the mixed number to an improper fraction. Then change the fraction to a decimal and then to a percent.

> **REMEMBER**
>
> Mixed numbers are fractions greater than 1. So the percents are always greater than 100%.

Do:

To change the mixed number to a percent:

Step 1. Change the mixed number into an improper fraction

$$2\frac{1}{2} = \frac{5}{2}$$

Step 2. Divide the numerator by the denominator

$$\begin{array}{r} 2.5 \\ 2\overline{)5.00} \\ \underline{4} \\ 10 \\ \underline{10} \end{array}$$

Step 3. Move the decimal point two places to the right.

250.

Step 4. Add the percent sign.

250%

Try These

Change the following mixed numbers to percents.

1. $2\frac{1}{2} =$ _____

2. $3\frac{1}{4} =$ _____

3. $10\frac{3}{4} =$ _____

49

4. $4\frac{3}{5} = $ _____ **5.** $1\frac{9}{10} = $ _____ **6.** $2\frac{2}{5} = $ _____

PRACTICE

Change the following mixed numbers to percents.

7. $1\frac{3}{10} = $ _____   **8.** $3\frac{13}{25} = $ _____   **9.** $2\frac{3}{5} = $ _____

10. $3\frac{7}{20} = $ _____   **11.** $4\frac{7}{10} = $ _____   **12.** $1\frac{1}{4} = $ _____

13. $1\frac{1}{10} = $ _____   **14.** $5\frac{1}{5} = $ _____   **15.** $2\frac{4}{25} = $ _____

16. $2\frac{9}{10} = $ _____   **17.** $6\frac{11}{20} = $ _____   **18.** $2\frac{12}{25} = $ _____

Solving Problems

Solve.

19. The big new discount supermarket will hire $3\frac{1}{2}$ times as many employees as the smaller supermarket that use to be there. Write $3\frac{1}{2}$ as a percent.

_____

20. In her print shop business, Samira has $2\frac{1}{4}$ times as many bids on new jobs as she did three years ago. Write $2\frac{1}{4}$ as a percent.

_____

Check your answers on page 66.

Louis has a budget to keep track of how much he spends. He has made a circle graph to show how he spends his money.

The circle is 100% of his earnings. Each section of the circle shows the percent he spends on each item. What percent does Louis spend for food?

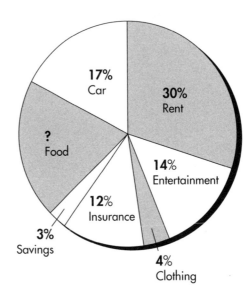

Think

Circle graphs help us see—and compare—the different parts of a whole. The circle graph above shows and compares the amounts that Louis spends on different things.

The parts of a 100% graph can be separated into percents, fractions, or decimals. The percents should always total 100%. The decimals or fractions in the graph should always add up to 1.

Do

The parts must add up to 100%. Add the parts you know then subtract the sum from 100%.

Step 1. 17% + 30% + 14% + 4% + 12% + 3% = 80%

Step 2. 100% − 80% = 20%

Louis spends 20% of his earnings on food.

Try These

Fill in the circle graph with the correct percents.

1. Sleeping   30%

    Working   40%

    Training   10%

    Studying   5%

    Leisure   3%

    Family   12%

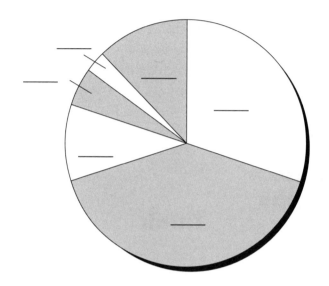

PRACTICE

2. Complete the circle graph using the following information. Leroy spends eight hours sleeping, seven hours working, one hour running, two hours watching television, three hours in his GED class, and three hours with his family. Assume that 24 hours is one whole.

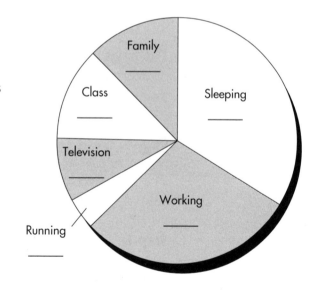

Check your answers on pages 66-67.

22

Applying
Your
Knowledge:
Percents,
Fractions,
and
Decimals

Try These

Match the percent in the middle column with the correct fraction and with the correct decimal.

	Fraction	Percent	Decimal
1.	.04	20%	$\frac{1}{4}$
2.	.2	60%	$\frac{3}{10}$
3.	.25	.2%	$\frac{4}{10}$
4.	.6	30%	$\frac{1}{25}$
5.	.002	4%	$\frac{1}{500}$
6.	.4	25%	$\frac{1}{5}$
7.	.3	.4%	$\frac{2}{500}$
8.	.004	40%	$\frac{3}{5}$

Complete the following by filling in the blanks with the equivalent fractions and equivalent decimals.

	Fraction	Percent	Decimal
9.	_____	15%	_____
10.	_____	90%	_____
11.	_____	5%	_____
12.	_____	1%	_____
13.	_____	75%	_____
14.	_____	50%	_____
15.	_____	8%	_____
16.	_____	2%	_____

Solving Problems

Solve.

17. A recent study has found that a new vaccine for one strain of the flu is 92% effective. Express this percent as a fraction and as a decimal.

_____

18. The approval rating for a new television show has recently jumped to 65%. Express this percent as a fraction and as a decimal.

_____

Check your answers on page 67.

23

Practical Percent Problems

To successfully solve problems with percent, you need to apply the different skills you have learned so far in this book. This lesson will help you review and practice these skills.

1. A newspaper reports that 65% of the people think the president is doing a good job. What fraction of the people approve of the president's performance? _____

2. Fifteen out of twenty people passed the driver's test on the first try. What percent of the people passed the test? _____

3. The sales tax is 7.5%. Write the decimal that would be used to figure the bill. _____

4. The credit card company used 0.129 to figure the interest charge on last month's purchases. What percent is being charged on this account? _____

5. Five out of six registered voters turned out for the last election. What fraction of the registered voters went to the polls? _____

6. The breakfast cereal has .15 of the recommended daily allowance of iron. What fraction of the daily iron is provided by the cereal? _____

7. Medical costs have increased by $1\frac{3}{5}$ in the past few years. By what percent have the college costs increased? _____

8. The school expenses of Joel's two children will increase 175% in the next two years because both of them are going on for more education after they graduate from high school. Change this percent to a fraction. _____

9. Imran and two of his friends were sharing a pizza for lunch. The pizza was cut into 8 pieces. If each person ate two pieces, write a fraction to express how much of the pizza they ate. Simplify the fraction. What percent of the pie does the fraction represent? _____

10. The interest rate on home loans is 12%. Express the interest rate as a decimal. _____

11. Hans sent surveys to 150 parents. So far, 120 surveys have been returned. What fraction of the parents returned the surveys? Simplify the fraction. _____

12. In his survey, Hans needs a response from at least 75% of the parents for his information to be valuable. Change the fraction from question 11 to a percent. Did Hans get a high enough percent of responses? _____

13. Hema's business expenses have increased by $\frac{2}{5}$ since last year. What percent would represent the increase? _____

14. Mr. Fuentes was reading about his favorite baseball player and saw that the player's batting average was .325. Change the batting average to a percent. _____

15. Evan's business has increased 225% since he opened his new fast-food franchise two years ago. What mixed number would represent the increase? _____

16. According to the local news, there is a 20% chance of rain today. What fraction would also represent the chance of rain? _____

17. Dave answered 35 out of 50 questions correctly on his math test. What percent of the questions did he answer correctly? _____

18. Only 9.5% of the workers failed to attend the meeting on the new flex-time hours. Change the percent to a decimal. _____

19. Write the percent that is shaded. Then write the percent as a fraction and as a decimal.

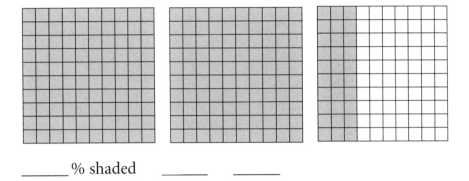

_____% shaded   _____   _____

20. Write the percent that is shaded. Then write the percent as a fraction and as a decimal.

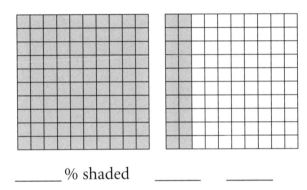

_____% shaded   _____   _____

21. Change each of the percents to a fraction.

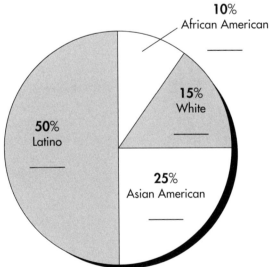

Ethnic Background of People Living in Rancho Viejo

22. Complete the circle graph. Assume that 24 hours is one whole.

Travel	1 hour
Misc.	4 hours
Sleep	6 hours
In class	3 hours
Eating	$1\frac{1}{2}$ hours
Work	8 hours
Dressing	$\frac{1}{2}$ hour

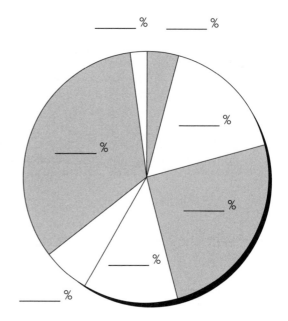

Check your answers on pages 67-68.

Answer Key

Lesson 1 ·······► FRACTIONS (PAGES 2–3)

1. $\dfrac{1}{3}$ 2. $\dfrac{5}{16}$ 3. $\dfrac{1}{6}$ 4. $\dfrac{3}{5}$ 5. $\dfrac{1}{4}$ 6. $\dfrac{5}{9}$

7. $\dfrac{5}{8}$ 8. $\dfrac{13}{15}$ 9. $\dfrac{75}{100}$ 10. $\dfrac{7}{7}$ 11. $\dfrac{2}{5}$ 12. $\dfrac{13}{100}$

Lesson 2 ·······► EQUIVALENT FRACTIONS (PAGES 5–7)

1. $\dfrac{6}{8}$ 2. $\dfrac{36}{40}$ 3. $\dfrac{8}{10}$ 4. $\dfrac{28}{32}$ 5. $\dfrac{3}{5}$ 6. $\dfrac{3}{4}$ 7. $\dfrac{8}{12}$ 8. $\dfrac{5}{6}$ 9. $\dfrac{12}{21}$

10. $\dfrac{1}{2}$ $\dfrac{50}{100} \div \dfrac{50}{50} = \dfrac{1}{2}$ 11. $\dfrac{3}{20}$ $\dfrac{15}{100} \div \dfrac{5}{5} = \dfrac{3}{20}$ 12. $\dfrac{3}{5}$ $\dfrac{6}{10} \div \dfrac{2}{2} = \dfrac{3}{5}$

13. $\dfrac{6}{25}$ $\dfrac{12}{50} \div \dfrac{2}{2} = \dfrac{6}{25}$ 14. $\dfrac{1}{10}$ $\dfrac{10}{100} \div \dfrac{10}{10} = \dfrac{1}{10}$ 15. $\dfrac{3}{4}$ $\dfrac{75}{100} \div \dfrac{25}{25} = \dfrac{3}{4}$

16. $\dfrac{25}{100}$ $\dfrac{1}{4} \times \dfrac{25}{25} = \dfrac{25}{100}$ 17. $\dfrac{6}{10}$ $\dfrac{3}{5} \times \dfrac{2}{2} = \dfrac{6}{10}$ 18. $\dfrac{35}{100}$ $\dfrac{7}{20} \times \dfrac{5}{5} = \dfrac{35}{100}$

19. $\dfrac{10}{100}$ $\dfrac{1}{10} \times \dfrac{10}{10} = \dfrac{10}{100}$ 20. $\dfrac{10}{60}$ $\dfrac{2}{12} \times \dfrac{5}{5} = \dfrac{10}{60}$ 21. $\dfrac{18}{100}$ $\dfrac{9}{50} \times \dfrac{2}{2} = \dfrac{18}{100}$

22. $\dfrac{2}{3}$ $\dfrac{50}{75} \div \dfrac{25}{25} = \dfrac{2}{3}$ 23. $\dfrac{8}{24} = \dfrac{1}{3}$ $\dfrac{8}{24} \div \dfrac{8}{8} = \dfrac{1}{3}$

Lesson 3 ·······► IMPROPER FRACTIONS AND MIXED NUMBERS (PAGE 9)

1. $\dfrac{28}{5}$ $5 \times 5 = 25$ $25 + 3 = 28$ 2. $\dfrac{5}{2}$ $2 \times 2 = 4$ $4 + 1 = 5$

3. $\dfrac{11}{3}$ $3 \times 3 = 9$ $9 + 2 = 11$ 4. $\dfrac{39}{8}$ $4 \times 8 = 32$ $32 + 7 = 39$

5. $\dfrac{29}{4}$ $7 \times 4 = 28$ $28 + 1 = 29$ 6. $\dfrac{25}{3}$ $8 \times 3 = 24$ $24 + 1 = 25$

7. $\dfrac{43}{8}$ $5 \times 8 = 40$ $40 + 3 = 43$ 8. $\dfrac{10}{7}$ $1 \times 7 = 7$ $7 + 3 = 10$

9. $2\dfrac{1}{3}$ $3\overline{)7}^{\,2\frac{1}{3}}$ 10. $4\dfrac{1}{2}$ $2\overline{)9}^{\,4\frac{1}{2}}$ 11. $3\dfrac{1}{3}$ $6\overline{)20}^{\,3\frac{2}{6}}$ $\dfrac{2}{6} \div \dfrac{2}{2} = \dfrac{1}{3}$ 12. $1\dfrac{2}{3}$ $3\overline{)5}^{\,1\frac{2}{3}}$

13. $1\dfrac{1}{2}$ $100\overline{)150}^{\,1\frac{50}{100}}$ $\dfrac{50}{100} \div \dfrac{50}{50} = \dfrac{1}{2}$ 14. $1\dfrac{1}{5}$ $10\overline{)12}^{\,1\frac{2}{10}}$ $\dfrac{2}{10} \div \dfrac{2}{2} = \dfrac{1}{5}$

15. $1\dfrac{1}{2}$ $50\overline{)75}^{\,1\frac{25}{50}}$ $\dfrac{25}{50} \div \dfrac{25}{25} = \dfrac{1}{2}$ 16. $1\dfrac{7}{8}$ $8\overline{)15}^{\,1\frac{7}{8}}$

Lesson 4 ·······▶ **DECIMALS** (PAGES 11–12)

1. .4   2. .05   3. .35   4. .8   5. .7   6. .45   7. .18   8. eighteen and eight hundredths

9. two and two tenths   10. one thousand, eight hundred fifty-six ten thousandths

11. thirty-five hundredths   12. four and ninety-one hundredths

13. 6.09   14. 501.4   15. 25.402   16. 1,056.22   17. 963.8

Lesson 5 ·······▶ **CHANGING FRACTIONS TO DECIMALS AND DECIMALS TO FRACTIONS** (PAGES 14–15)

1. .12   2. .25   3. .7   4. .009   5. .6 $5\overline{)3.0}^{\,.6}$

6. .65 $20\overline{)13.00}^{\,.65}$ 7. .57 $7\overline{)4.000}^{\,.571}$ 8. .63 $8\overline{)5.000}^{\,.625}$ 9. .8 $5\overline{)4.0}^{\,.8}$ 10. .83 $6\overline{)5.00}^{\,.83}$

11. $\dfrac{3}{5}$ $\dfrac{6}{10} \div \dfrac{2}{2} = \dfrac{3}{5}$ 12. $\dfrac{9}{20}$ $\dfrac{45}{100} \div \dfrac{5}{5} = \dfrac{9}{20}$ 13. $\dfrac{1}{4}$ $\dfrac{25}{100} \div \dfrac{25}{25} = \dfrac{1}{4}$

14. $\dfrac{2}{25}$ $\dfrac{8}{100} \div \dfrac{4}{4} = \dfrac{2}{25}$ 15. $\dfrac{3}{25}$ $\dfrac{12}{100} \div \dfrac{4}{4} = \dfrac{3}{25}$ 16. $\dfrac{4}{5}$ $\dfrac{80}{100} \div \dfrac{20}{20} = \dfrac{4}{5}$

17. 1. 15 lbs $4\overline{)3.00}^{\,.75}$ $5\overline{)2.0}^{\,.4}$ .75 + .4 = 1.15   18. .75   .5 $4\overline{)3.00}^{\,.75}$ $2\overline{)1.0}^{\,.5}$

Lesson 6 ·······▶ **APPLYING YOUR KNOWLEDGE: FRACTIONS AND DECIMALS** (PAGES 16–17)

1. .75 $\dfrac{3}{4}$ 2. .26 $\dfrac{13}{50}$ 3. .2 $\dfrac{1}{5}$

4. .25 $\dfrac{1}{4}$ 5. .3 $\dfrac{3}{10}$ 6. $\dfrac{1}{5}$ .2

7. $\dfrac{2}{100}$ .02   8. $\dfrac{4}{100}$ .04   9. $\dfrac{5}{10}$ .5

10. $\dfrac{1}{4}$ .25   11. $\dfrac{3}{6}$,   .5, $\dfrac{1}{2}$

Lesson 7 ·····► WHAT IS PERCENT? (PAGES 18-19)

1. 10% **2.** 25% **3.** 1% **4.** 3% **5.** 5% **6.** 97% **7.** 40% **8.** 18% **9.** 15%

10.   **11.**   **12.**

13.   **14.**   **15.**

Lesson 8 ·····► PERCENTS GREATER THAN 100% (PAGES 21-22)

1. 105% **2.** 195% **3.** 115% **4.** 102% **5.** 128% **6.** 199%

7. 200% **8.** 125% **9.** 160% **10.**

11.  **12.** 200% **13.** 150% **14.**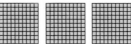

15. This means that the price of beef increased 50% over the original price.

Lesson 9 ·····► APPLYING YOUR KNOWLEDGE: MEANING OF PERCENT (PAGES 23-25)

1. "for each 100" or "out of 100" **2.** fraction or decimal

3. 25% shaded   75% not shaded **4.** 90% shaded   10% not shaded

5. 60% shaded   40% not shaded **6.** 12% shaded   88% not shaded

7. 55% shaded   45% not shaded **8.** 92% shaded   8% not shaded

9. 175% **10.** 105% **11.** 195% **12.** 115% **13.** 125% **14.** 160%

15. \$2 **16.**   **17.** $\frac{1}{20}$ **18.**   **19.** 28% **20.** none of them

Lesson 10 ·····► CHANGING PERCENTS TO FRACTIONS (PAGES 26-27)

1. $\frac{17}{100}$

2. $\frac{4}{5}$ $\frac{80}{100} \div \frac{20}{20} = \frac{4}{5}$

3. $\frac{1}{5}$ $\frac{20}{100} \div \frac{20}{20} = \frac{1}{5}$

4. $\frac{8}{25}$ $\frac{32}{100} \div \frac{4}{4} = \frac{8}{25}$

5. $\frac{1}{2}$ $\frac{50}{100} \div \frac{50}{50} = \frac{1}{2}$

6. $\frac{7}{25}$ $\frac{28}{100} \div \frac{4}{4} = \frac{7}{25}$

7. $\dfrac{7}{20}$ $\qquad$ $\dfrac{35}{100} \div \dfrac{5}{5} = \dfrac{7}{20}$

8. $\dfrac{17}{25}$ $\qquad$ $\dfrac{68}{100} \div \dfrac{4}{4} = \dfrac{17}{25}$

9. $\dfrac{3}{20}$ $\qquad$ $\dfrac{15}{100} \div \dfrac{5}{5} = \dfrac{3}{20}$

10. $\dfrac{14}{25}$ $\qquad$ $\dfrac{56}{100} \div \dfrac{4}{4} = \dfrac{14}{25}$

11. $\dfrac{39}{50}$ $\qquad$ $\dfrac{78}{100} \div \dfrac{2}{2} = \dfrac{39}{50}$

12. $\dfrac{31}{50}$ $\qquad$ $\dfrac{62}{100} \div \dfrac{2}{2} = \dfrac{31}{50}$

13. $\dfrac{9}{10}$ $\qquad$ $\dfrac{90}{100} \div \dfrac{10}{10} = \dfrac{9}{10}$

14. $\dfrac{1}{20}$ $\qquad$ $\dfrac{5}{100} \div \dfrac{5}{5} = \dfrac{1}{20}$

15. $\dfrac{12}{25}$ $\qquad$ $\dfrac{48}{100} \div \dfrac{4}{4} = \dfrac{12}{25}$

16. $\dfrac{21}{25}$ $\qquad$ $\dfrac{84}{100} \div \dfrac{4}{4} = \dfrac{21}{25}$

17. $\dfrac{4}{5}$ $\qquad$ $\dfrac{80}{100} \div \dfrac{20}{20} = \dfrac{4}{5}$

18. $\dfrac{13}{20}$ $\qquad$ $\dfrac{65}{100} \div \dfrac{5}{5} = \dfrac{13}{20}$

Lesson 11 ······▶ CHANGING FRACTIONS TO PERCENTS (PAGES 28–29)

1. $\dfrac{1}{2} = 50\%$ $\qquad$ $2\overline{)1.00}^{\,.50}$

2. $\dfrac{7}{10} = 70\%$ $\qquad$ $10\overline{)7.00}^{\,.70}$

3. $\dfrac{9}{20} = 45\%$ $\qquad$ $20\overline{)9.00}^{\,.45}$

4. $\dfrac{1}{4} = 25\%$ $\qquad$ $4\overline{)1.00}^{\,.25}$

5. $\dfrac{7}{20} = 35\%$ $\qquad$ $20\overline{)7.00}^{\,.35}$

6. $\dfrac{3}{5} = 60\%$ $\qquad$ $5\overline{)3.00}^{\,.60}$

7. $\dfrac{1}{8} = 13\%$ $\qquad$ $8\overline{)1.000}^{\,.125}$

8. $\dfrac{5}{6} = 83\%$ $\qquad$ $6\overline{)5.000}^{\,.833}$

9. $\dfrac{11}{20} = 55\%$ $\qquad$ $20\overline{)11.00}^{\,.55}$

10. $\dfrac{2}{3} = 67\%$ $\qquad$ $3\overline{)2.000}^{\,.666}$

11. $\dfrac{17}{100} = 17\%$ $\qquad$ $100\overline{)17.00}^{\,.17}$

12. $\dfrac{7}{9} = 78\%$ $\qquad$ $9\overline{)7.000}^{\,.777}$

13. $\dfrac{8}{10} = 80\%$ $\qquad$ $10\overline{)8.00}^{\,.80}$

14. $\dfrac{9}{25} = 36\%$ $\qquad$ $25\overline{)9.00}^{\,.36}$

15. 65% $\qquad$ $20\overline{)13.00}^{\,.65}$

16. 15% $\qquad$ $20\overline{)3.00}^{\,.15}$

Lesson 12 ······▶ APPLYING YOUR KNOWLEDGE: PERCENTS AND FRACTIONS (PAGES 30–32)

1. $\dfrac{7}{20}$ $\quad$ 35%

2. $\dfrac{1}{5}$ $\quad$ 20%

3. $\dfrac{3}{10}$ $\quad$ 30%

4. $\dfrac{1}{2}$ $\quad$ 50%

5. $\dfrac{3}{20}$ $\quad$ 15%

6. $\dfrac{3}{4}$ $\quad$ 75%

7. $\frac{3}{5}$ 60%

8. $\frac{9}{10}$ 90%

9. $\frac{1}{4}$ 25%

10. $\frac{13}{20}$ 65%

11. $\frac{2}{5}$ 40%

12. $\frac{4}{5}$ 80%

13. $\frac{1}{10}$ 10%

14. $\frac{9}{20}$ 45%

15. $\frac{1}{2}$ 50%

16. $\frac{7}{10}$ 70%

Lesson 13 ·······▶ 50% AND 25% (PAGE 34)

1.

2. (diagram)

3. (diagram)

4.

5.

6. (diagram)

Lesson 14 ·······▶ COMMON PERCENTS AND FRACTION EQUIVALENTS
(PAGES 36–37)

1. $\frac{1}{2}$

2. 20%

3. $\frac{1}{8}$

4. $33\frac{1}{3}\%$

5. $\frac{1}{10}$

6. 75%

7. $\frac{1}{20}$

8. 25%

9. $\frac{1}{3}$

10. 50%

11. $\frac{1}{4}$

12. 10%

13. $\frac{2}{3}$

14. $12\frac{1}{2}\%$

15. $\frac{3}{4}$

16. 5%

17. $\frac{1}{5}$

18. $66\frac{2}{3}\%$

19. $\frac{1}{2}$ $50\% = \frac{50}{100} \div \frac{50}{50} = \frac{1}{2}$

20. 60% $\frac{3}{5} = \begin{array}{r} .60 \\ 5\overline{)3.00} \end{array}$

21. $\frac{4}{5}$ $80\% = \frac{80}{100} \div \frac{20}{20} = \frac{4}{5}$

22. Ernesto pays 40%. $\frac{2}{5} = \begin{array}{r} .40 \\ 5\overline{)2.00} \end{array}$
    Mr. Rivera pays 60%.
    $100\% - 40\% = 60\%$

23. 30% $\frac{3}{10} = \begin{array}{r} .30 \\ 10\overline{)3.00} \end{array}$

24. $\frac{1}{20}$ $5\% = \frac{5}{100} \div \frac{5}{5} = \frac{1}{20}$

Lesson 15 ·····▶ CHANGING PERCENTS TO DECIMALS (PAGES 38–39)

1. .34   **2.** .72   **3.** .08   **4.** .03   **5.** .30   **6.** .80   **7.** .076   **8.** .457

9. .855   **10.** .03   **11.** .105   **12.** .044   **13.** .05   **14.** .0725   **15.** .07   **16.** .085

Lesson 16 ·····▶ CHANGING DECIMALS TO PERCENTS (PAGES 40–41)

1. 63%   **2.** 51%   **3.** 20%   **4.** 46.5%   **5.** 18%   **6.** 80%

7. 49%   **8.** 5%   **9.** 0.5%   **10.** 60%   **11.** 28.1%   **12.** 90%

13. 33%   **14.** 87.6%   **15.** 20%   **16.** 44.4%   **17.** 30.8%   **18.** 15%

Lesson 17 ·····▶ PERCENTS LESS THAN 1% (PAGES 42–43)

1. .003 $\dfrac{3}{1,000}$ $3\% = .003 = \dfrac{3}{1,000}$ **2.** .004 $\dfrac{1}{250}$ $.4\% = .004 = \dfrac{4}{1,000} \div \dfrac{4}{4} = \dfrac{1}{250}$

3. .0025 $\dfrac{1}{400}$ $.25\% = .0025 = \dfrac{25}{10,000} \div \dfrac{25}{25} = \dfrac{1}{400}$

4. .002 $\dfrac{1}{500}$ $.2\% = .002 = \dfrac{2}{1000} \div \dfrac{2}{2} = \dfrac{1}{500}$

5. .008   **6.** .075   **7.** .0015   **8.** .00098   **9.** .0315   **10.** .0054

11. $\dfrac{2}{625}$ $.32\% = .0032 = \dfrac{32}{10,000} \div \dfrac{16}{16} = \dfrac{2}{625}$

12. $\dfrac{4}{625}$ $.64\% = .0064 = \dfrac{64}{10,000} \div \dfrac{16}{16} = \dfrac{4}{625}$

13. $\dfrac{1}{40}$ $2.5\% = .025 = \dfrac{25}{1,000} \div \dfrac{25}{25} = \dfrac{1}{40}$

14. $\dfrac{3}{500}$ $.6\% = .006 = \dfrac{6}{1,000} \div \dfrac{2}{2} = \dfrac{3}{500}$

15. $\dfrac{3}{625}$ $.48\% = .0048 = \dfrac{48}{10,000} \div \dfrac{16}{16} = \dfrac{3}{625}$

16. 33%   **17.** .9%   **18.** .75% $\dfrac{3}{400} = 400\overline{)3.0000}^{\,.0075}$ **19.** 1.2% $\dfrac{3}{250} = 250\overline{)3.000}^{\,.012}$

20. .39%   **21.** 7.05%   **22.** .0004 $\dfrac{1}{2,500}$ $.04\% = .0004 = \dfrac{4}{10,000} \div \dfrac{4}{4} = \dfrac{1}{2,500}$

Lesson 18 ┈┈► DECIMALS AND PERCENTS GREATER THAN 100% (PAGES 44-46)

1. 2.25	**2.** 1.33	**3.** 3.15	**4.** 1.85	**5.** 3.78	**6.** 2.05	**7.** 3.55
8. 4.50	**9.** 1.25	**10.** 155%	**11.** 285%	**12.** 235%	**13.** 320%	**14.** 810%
15. 305%	**16.** 405%	**17.** 215%	**18.** 601%	**19.** 2.56	**20.** 105%	**21.** 1.96
22. 290%	**23.** 6.51	**24.** 375%	**25.** 7.58	**26.** 462%	**27.** 1.05	**28.** 533%
29. 2.47	**30.** 108%	**31.** 2.25	**32.** 1.25	**33.** 3.00	**34.** 2.20	**35.** 1.15

Lesson 19 ┈┈► CHANGING PERCENTS GREATER THAN 100% TO MIXED NUMBERS (PAGES 47-48)

1. $2\frac{1}{2}$ $250\% = \frac{250}{100} \div \frac{50}{50} = \frac{5}{2} = 2\frac{1}{2}$

2. $1\frac{1}{4}$ $125\% = \frac{125}{100} \div \frac{25}{25} = \frac{5}{4} = 1\frac{1}{4}$

3. $3\frac{3}{4}$ $375\% = \frac{375}{100} \div \frac{25}{25} = \frac{15}{4} = 3\frac{3}{4}$

4. $2\frac{1}{5}$ $220\% = \frac{220}{100} \div \frac{20}{20} = \frac{11}{5} = 2\frac{1}{5}$

5. $1\frac{4}{5}$ $180\% = \frac{180}{100} \div \frac{2}{2} = \frac{90}{50} = 1\frac{4}{5}$

6. $4\frac{1}{4}$ $425\% = \frac{425}{100} \div \frac{25}{25} = \frac{17}{4} = 4\frac{1}{4}$

7. $3\frac{3}{10}$ $330\% = \frac{330}{100} \div \frac{2}{2} = \frac{165}{50} = 3\frac{3}{10}$

8. $1\frac{3}{20}$ $115\% = \frac{115}{100} \div \frac{5}{5} = \frac{23}{20} = 1\frac{3}{20}$

9. $3\frac{17}{20}$ $385\% = \frac{385}{100} \div \frac{5}{5} = \frac{77}{20} = 3\frac{17}{20}$

10. $2\frac{3}{20}$ $215\% = \frac{215}{100} \div \frac{5}{5} = \frac{43}{20} = 2\frac{3}{20}$

11. $1\frac{9}{20}$ $145\% = \frac{145}{100} \div \frac{5}{5} = \frac{29}{20} = 1\frac{9}{20}$

12. $\frac{3}{5}$ $60\% = \frac{60}{100} = \frac{3}{5}$

13. $4\frac{4}{5}$ $480\% = \frac{480}{100} \div \frac{20}{20} = \frac{24}{5} = 4\frac{4}{5}$

14. $1\frac{1}{20}$ $105\% = \frac{105}{100} \div \frac{5}{5} = \frac{21}{20} = 1\frac{1}{20}$

15. $\frac{1}{20}$ $5\% = \frac{5}{100} \div \frac{5}{5} = \frac{1}{20}$

16. $2\frac{4}{5}$ $280\% = \frac{280}{100} \div \frac{20}{20} = \frac{14}{5} = 2\frac{4}{5}$

17. $\frac{3}{10}$ $30\% = \frac{30}{100} \div \frac{10}{10} = \frac{3}{10}$

18. $3\frac{3}{5}$ $360\% = \frac{360}{100} \div \frac{4}{4} = \frac{90}{25} = 3\frac{15}{25} = 3\frac{3}{5}$

19. $3\frac{9}{10}$ $390\% = \frac{390}{100} \div \frac{10}{10} = \frac{39}{10} = 3\frac{9}{10}$

20. $1\frac{11}{20}$ $155\% = \frac{155}{100} \div \frac{5}{5} = \frac{31}{20} = 1\frac{11}{20}$

21. $1\frac{3}{10}$ $130\% = \frac{130}{100} \div \frac{10}{10} = \frac{13}{10} = 1\frac{3}{10}$

22. $2\frac{1}{4}$ $225\% = \frac{225}{100} \div \frac{25}{25} = \frac{9}{4} = 2\frac{1}{4}$

Lesson 20 ·······▶ CHANGING MIXED NUMBERS TO PERCENTS
(PAGES 49-50)

1. 250% $2\dfrac{1}{2} = \dfrac{5}{2} = 2\overline{)5.00}\,^{2.50}$

2. 325% $3\dfrac{1}{4} = \dfrac{13}{4} = 4\overline{)13.00}\,^{3.25}$

3. 1075% $10\dfrac{3}{4} = \dfrac{43}{4} = 4\overline{)43.00}\,^{10.75}$

4. 460% $4\dfrac{3}{5} = \dfrac{23}{5} = 5\overline{)23.00}\,^{4.60}$

5. 190% $1\dfrac{9}{10} = \dfrac{19}{10} = 10\overline{)19.00}\,^{1.90}$

6. 240% $2\dfrac{2}{5} = \dfrac{12}{5} = 5\overline{)12.00}\,^{2.40}$

7. 130% $1\dfrac{3}{10} = \dfrac{13}{10} = 10\overline{)13.00}\,^{1.30}$

8. 352% $3\dfrac{13}{25} = \dfrac{88}{25} = 25\overline{)88.00}\,^{3.52}$

9. 260% $2\dfrac{3}{5} = \dfrac{13}{5} = 5\overline{)13.00}\,^{2.60}$

10. 335% $3\dfrac{7}{20} = \dfrac{67}{20} = 20\overline{)67.00}\,^{3.35}$

11. 470% $4\dfrac{7}{10} = \dfrac{47}{10} = 10\overline{)47.00}\,^{4.70}$

12. 125% $1\dfrac{1}{4} = \dfrac{5}{4} = 4\overline{)5.00}\,^{1.25}$

13. 110% $1\dfrac{1}{10} = \dfrac{11}{10} = 10\overline{)11.00}\,^{1.10}$

14. 520% $5\dfrac{1}{5} = \dfrac{26}{5} = 5\overline{)26.00}\,^{5.20}$

15. 216% $2\dfrac{4}{25} = \dfrac{54}{25} = 25\overline{)54.00}\,^{2.16}$

16. 290% $2\dfrac{9}{10} = \dfrac{29}{10} = 10\overline{)29.00}\,^{2.90}$

17. 655% $6\dfrac{11}{20} = \dfrac{131}{20} = 20\overline{)131.00}\,^{6.55}$

18. 248% $2\dfrac{12}{25} = \dfrac{62}{25} = 25\overline{)62.00}\,^{2.48}$

19. 350% $3\dfrac{1}{2} = \dfrac{7}{2} = 2\overline{)7.00}\,^{3.50}$

20. 225% $2\dfrac{1}{4} = \dfrac{9}{4} = 4\overline{)9.00}\,^{2.25}$

Lesson 21 ·······▶ CIRCLE GRAPHS   (PAGE 52)

1.

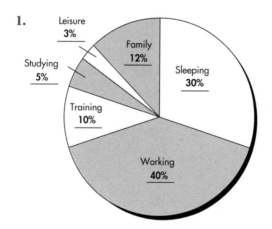

2.

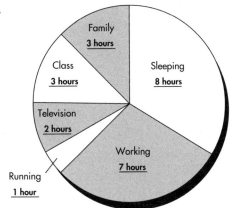

Lesson 22 ·······▶ **APPLYING YOUR KNOWLEDGE: PERCENTS, FRACTIONS, AND DECIMALS** (PAGES 53–54)

1. .2   20% $\frac{1}{5}$ **2.** .6   60% $\frac{3}{5}$ **3.** .002   .2% $\frac{1}{500}$ **4.** .3   30% $\frac{3}{10}$

5. .04   4% $\frac{1}{25}$ **6.** .25   25% $\frac{1}{4}$ **7.** .004   .4% $\frac{2}{500}$ **8.** .4   40% $\frac{4}{10}$

9. $\frac{3}{20}$ 15%   .15   **10.** $\frac{9}{10}$ 90%   .9   **11.** $\frac{1}{20}$ 5%   .05   **12.** $\frac{1}{100}$ 1%   .01

13. $\frac{3}{4}$ 75%   .75   **14.** $\frac{1}{2}$ 50%   .5   **15.** $\frac{2}{25}$ 8%   .08   **16.** $\frac{1}{50}$ 2%   .02

17. $\frac{23}{25}$ .92 $92\% = .92 = \frac{92}{100} \div \frac{4}{4} = \frac{23}{25}$ **18.** $\frac{13}{20}$ .65 $65\% = .65 = \frac{65}{100} \div \frac{5}{5} = \frac{13}{20}$

Lesson 23 ·······▶ **PRACTICAL PERCENT PROBLEMS** (PAGES 55–58)

1. $\frac{13}{20}$ $65\% = \frac{65}{100} \div \frac{5}{5} = \frac{13}{20}$ **2.** 75% $\frac{15}{20} = 20\overline{)15.00}^{.75}$ **3.** .075   **4.** 12.9%

5. $\frac{5}{6}$ **6.** $\frac{3}{20}$ $.15 = \frac{15}{100} \div \frac{5}{5} = \frac{3}{20}$ **7.** 160% $1\frac{3}{5} = \frac{8}{5} = 5\overline{)8.00}^{1.60}$

8. $1\frac{3}{4}$ $175\% = \frac{175}{100} \div \frac{25}{25} = \frac{7}{4} = 1\frac{3}{4}$ **9.** $\frac{3}{4}$ 75% $\frac{6}{8} \div \frac{2}{2} = \frac{3}{4}$ $4\overline{)3.00}^{.75}$ **10.** .12

11. $\frac{4}{5}$ $\frac{120}{150} \div \frac{30}{30} = \frac{4}{5}$ **12.** 80%   Yes $\frac{4}{5} = 5\overline{)4.00}^{.80}$ **13.** 40% $\frac{2}{5} = 5\overline{)2.00}^{.40}$

14. 32.5%   **15.** $2\frac{1}{4}$ $225\% = \frac{225}{100} \div \frac{25}{25} = \frac{9}{4} = 2\frac{1}{4}$ **16.** $\frac{1}{5}$ $20\% = \frac{20}{100} = \frac{1}{5}$

17. 70% $\quad \dfrac{35}{50} = 50\overline{)35.00}^{\,.70}$ **18.** .095

19. 230% $\quad 2\dfrac{3}{10} \quad 2.3 \quad \dfrac{230}{100} \div \dfrac{10}{10} = \dfrac{23}{10} = \dfrac{20}{10} + \dfrac{3}{10} = 2\dfrac{3}{10} \quad 10\overline{)23.0}^{\,2.3}$

20. 120% $\quad 1\dfrac{1}{5} \quad 1.2 \quad \dfrac{120}{100} \div \dfrac{20}{20} = \dfrac{6}{5} = \dfrac{5}{5} + \dfrac{1}{5} = 1\dfrac{1}{5} \quad 5\overline{)6.00}^{\,1.2}$

21. $\frac{1}{2}$ = 50% Latino $\qquad \dfrac{150}{100} \div \dfrac{50}{50} = \dfrac{1}{2}$

$\frac{1}{4}$ = 25% Asian $\qquad \dfrac{25}{100} \div \dfrac{25}{25} = \dfrac{1}{4}$

$\frac{3}{20}$ = 15% White $\qquad \dfrac{15}{100} \div \dfrac{5}{5} = \dfrac{3}{20}$

$\frac{1}{10}$ = 10% African American $\qquad \dfrac{10}{100} \div \dfrac{10}{10} = \dfrac{1}{10}$

22.

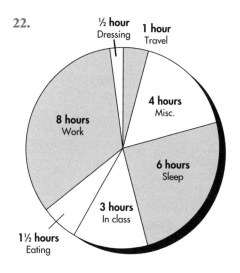

NOTES

NOTES

NOTES

NOTES

NOTES

NOTES